WHY SHOULD I EAT THIS?

WATER

TABLE OF CONTENTS

By Lindsay Snea

A Crabtree Seedlings Book

Crabtree Publishing

crabtreebooks.com

School-to-Home Support for Caregivers and Teachers

This book helps children grow by letting them practice reading. Here are a few guiding questions to help the reader with building his or her comprehension skills. Possible answers appear here in red.

Before Reading:

- What do I think this book is about?
  - *I think this book is about water.*
  - *I think this book is about why we need water.*
- What do I want to learn about water?
  - *I want to learn why drinking water is important.*
  - *I want to learn how much water we should drink every day.*

During Reading:

- I wonder why...
  - *I wonder why we should drink more water after exercising. I wonder why being dehydrated can make us feel tired.*
- What have I learned so far?
  - *I have learned that more than half of the body is made of water.*
  - *I have learned that drinking water is good for our muscles.*

After Reading:

- What details did I learn about water?
  - *I have learned that we should drink five to eight glasses of water a day.*
  - *I have learned that we can get water from some of the foods that we eat.*
- Read the book again and look for the vocabulary words.
  - *I see the word **energy** on page 9, and the word **dehydrated** on page 14. The other glossary words are found on pages 22 and 23.*

WATER

Drinking water is one of the most important things you can do for your body.

Water keeps you healthy. Without water, your body does not work the way it should.

More than half of your body is made up of water. So it is important to drink lots of water every day.

FACT

*More than two-thirds of Earth's **surface** is covered in water in the form of oceans, lakes, and rivers.*

The blood in your body is made mostly of water. Without water, your blood would not flow.

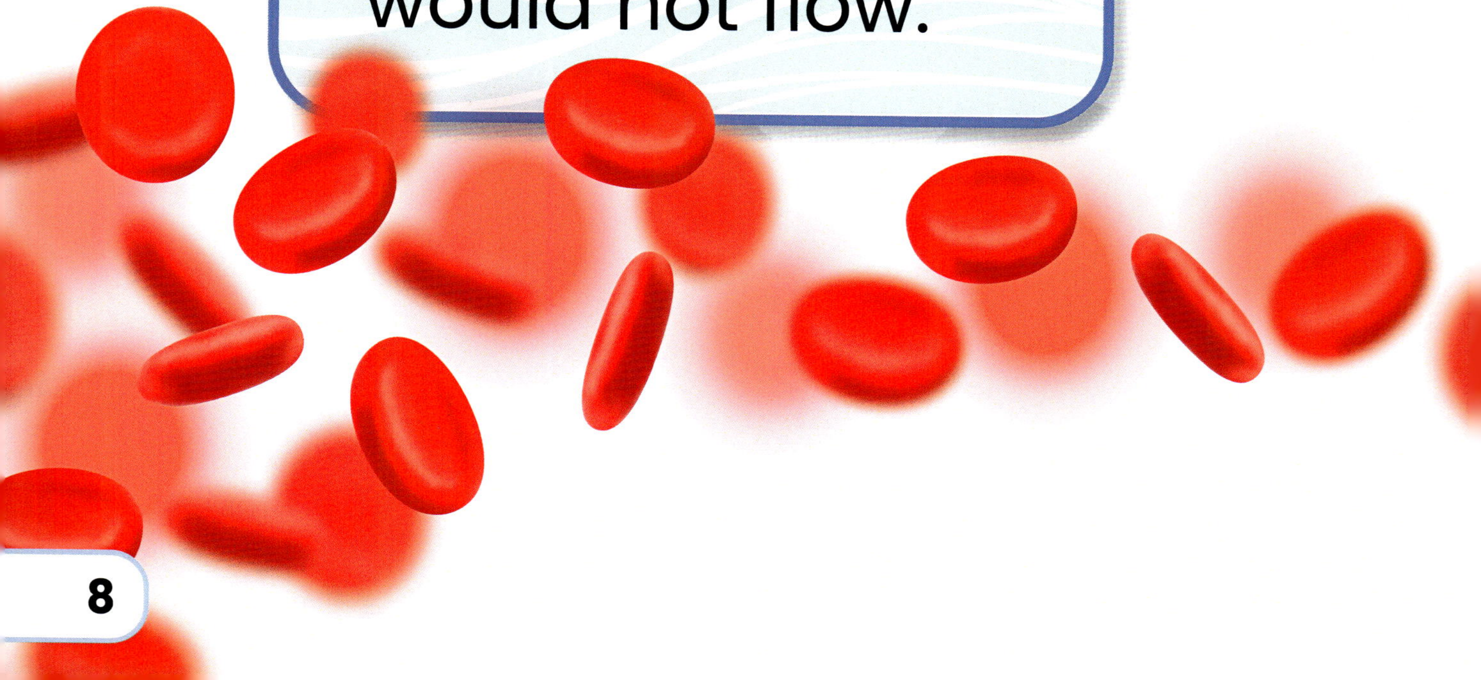

DID YOU KNOW?

*Blood brings **energy** to all the parts of your body. Your body needs energy to play, to learn, and to grow.*

Water helps your body break down and use **vitamins** and other important **nutrients** from the foods you eat.

Drinking water also helps keep your muscles, bones, teeth, and skin healthy.

Your body becomes **dehydrated** if you do not drink enough water. This can make you feel tired, grumpy, or dizzy.

You should try to drink five to eight glasses of water each day. It is good to drink more water on hot summer days or after exercising.

You can also get water from some of the foods you eat. Fruits and vegetables such as watermelons, cucumbers, and celery contain a lot of water.

It is healthier to drink water instead of sugary sodas when you are thirsty.

Glossary

dehydrated (dee-HAHY-dreyt-ed): Not having enough water in your body

energy (EN-er-jee): Your body or mind's ability to be active or do work

nutrients (NOO-tree-uhnts): Substances that your body needs to live and grow

surface (SUR-fis): The top layer of an area of land

vitamins (VAHY-tuh-mins): Nutrients found in foods that help your body grow and stay healthy

Index

About the Author

Lindsay Snea lives in the mountains of Fernie, British Columbia, Canada, with her husband and four children. Lindsay is an elementary school teacher, and a lifelong learner, who is passionate about cooking, food, fitness, being outdoors, and health.

Written by: lindsay Snea
Designed by: Margaret Salter
Proofreader: Melissa Boyce
Photographs: All images from Shutterstock
Production manager: Candice Campbell

Crabtree Publishing

crabtreebooks.com 800-387-7650

 In Canada: We acknowledge the financial support of the Government of Canada through the Canada Book Fund for our publishing activities.

Printed in the U.S.A./CP112025

Published in Canada
Crabtree Publishing
616 Welland Avenue
St. Catharines, Ontario
L2M 5V6

Published in the United States
Crabtree Publishing
347 Fifth Avenue
Suite 1402-145
New York, New York, 10016

Library and Archives Canada Cataloguing in Publication
Available at Library and Archives Canada

Library of Congress Cataloging-in-Publication Data
Available at the Library of Congress

Hardcover: 978-1-0398-8115-0
Paperback: 978-1-0398-8476-2
Ebook (pdf): 978-1-0398-8235-5
Epub: 978-1-0398-8355-0
Read-Along: 978-1-0398-8560-8
Audio: 978-1-0398-8518-9